Copyright Information

I0391149

June 13, 2016

Dear Colleagues:

Commercial and military aviation are fundamental drivers of national prosperity, mobility, and security. Yet the aviation industry faces significant energy and environmental challenges due to dependence on petroleum jet fuels. Liquid "drop-in" jet fuels derived from nonpetroleum feedstock that can replace conventional petroleum-based jet fuel without the need to modify aircraft engines and the fuel distribution infrastructure can help address these challenges. Development of these drop-in alternative jet fuels (AJFs) and other fuel alternatives facilitates a diverse, secure, and reliable fuel supply, and their use enhances U.S. energy security, contributes to jet fuel's price stability, reduces emissions that impact air quality and global climate, and generates rural economic development.

Among transportation fuel users, aviation is uniquely positioned for industry-wide use of AJFs. Unlike cars, planes have no near-term alternatives to liquid fuels, and they benefit from a concentrated fueling infrastructure with a limited number of fueling stations (airports) and a limited number of large fuel buyers (airlines and military). Aviation benefits from an aligned industry and government approach to fuel development and approval. Further, at the 37th United Nations International Civil Aviation Organization Assembly, the industry and member nations, including the United States, agreed to achieve carbon neutral aviation growth beginning in 2020, for which AJFs may prove to be crucial.

Over the past decade, significant progress has been made by commercial and military aviation to develop, evaluate, and deploy AJFs that can cost-effectively meet the challenges described above. Since 2009 ASTM International has approved five different types of AJFs. The past year has witnessed more than a half dozen announcements in the United States of fuel purchase agreements between renewable fuel producers, airlines, and the military. But at present, AJFs that compete with petroleum fuel on price are not yet produced in volumes sufficient to meet the needs of the aviation industry.

This *Federal Alternative Jet Fuels Research and Development Strategy* sets out prioritized Federal R&D goals and objectives to address key scientific and technical challenges that inhibit the development, production, and use of economically viable AJFs at commercial scale. It was developed with input from stakeholders under the auspices of the Aeronautics Science and Technology Subcommittee (ASTS) of the National Science and Technology Council by the Alternative Jet Fuel Interagency Working Group, which is made up of expert representatives from across the Federal Government. It aligns Federal agency research and development efforts to contribute to the successful mobilization of Federal and non-Federal stakeholder communities towards a common effort to realize the great promise presented by AJFs.

Jaiwon Shin
Co-Chair, ASTS
Associate Administrator
NASA Aeronautics Research
Mission Directorate

Spiro G. Lekoudis
Co-Chair, ASTS
Director, Weapons Systems,
Defense Research and
Engineering, Office of the
Secretary of Defense

Shelley Yak
Co-Chair, ASTS
Director, William J. Hughes
Technical Center, Federal
Aviation Administration

About the National Science and Technology Council

The National Science and Technology Council (NSTC) is the principal means by which the Executive Branch coordinates science and technology policy across the diverse entities that make up the Federal research and development (R&D) enterprise. One of the NSTC's primary objectives is establishing clear national goals for Federal science and technology investments. The NSTC prepares R&D packages aimed at accomplishing multiple national goals. The NSTC's work is organized under five committees: Environment, Natural Resources, and Sustainability; Homeland and National Security; Science, Technology, Engineering, and Mathematics (STEM) Education; Science; and Technology. Each of these committees oversees subcommittees and working groups that are focused on different aspects of science and technology. More information is available at www.whitehouse.gov/ostp/nstc.

About the Office of Science and Technology Policy

The Office of Science and Technology Policy (OSTP) was established by the National Science and Technology Policy, Organization, and Priorities Act of 1976. OSTP's responsibilities include advising the President in policy formulation and budget development on questions in which science and technology are important elements; articulating the President's science and technology policy and programs; and fostering strong partnerships among Federal, state, and local governments, and the scientific communities in industry and academia. The Director of OSTP also serves as Assistant to the President for Science and Technology and manages the NSTC. More information is available at www.whitehouse.gov/ostp.

About the Aeronautics Science and Technology Subcommittee

The Aeronautics Science and Technology Subcommittee (ASTS) of the NSTC Committee on Technology was established to advise and assist the Committee developing and advancing policies, strategies, and plans relating to federally sponsored aeronautics research and development to achieve broad national goals and support innovation. ASTS has functioned as a coordinating body to facilitate aeronautics R&D efforts across the Federal Government. The ASTS was instrumental in developing the *National Aeronautics Research and Development Policy* and the *National Plan for Aeronautics Research and Development and Related Infrastructure*.

About the Alternative Jet Fuel Interagency Working Group

The ASTS established the Alternative Jet Fuel Interagency Working Group (AJF-IWG) to assess Federal efforts to enable "new aviation fuels derived from diverse and domestic resources to improve fuel supply and price stability," a goal laid out in the 2010 *National Aeronautics Research and Development Plan*.

Acknowledgements

The AJF-IWG appreciates the analytical contributions to this effort by the following current and former individuals of the IDA Science and Technology Policy Institute (STPI): Emily J. Sylak-Glassman, Bhavya Lal, and Lucas M. Pratt. In addition, John C. Everett, Linda S. Garlet, and Erika T. Tildon of STPI provided editorial and production assistance.

About this Document

This document was developed by the AJF-IWG, and published by OSTP.

Report prepared by

NATIONAL SCIENCE AND TECHNOLOGY COUNCIL
COMMITTEE ON TECHNOLOGY
AERONAUTICS SCIENCE AND TECHNOLOGY SUBCOMMITTEE
ALTERNATIVE JET FUEL INTERAGENCY WORKING GROUP

National Science and Technology Council

Chair
John P. Holdren
Assistant to the President for Science
and Technology and Director,
Office of Science and Technology Policy

Staff
Afua Bruce
Executive Director

Committee on Technology

Chair
Thomas Kalil
Deputy Director, Technology and Innovation
Office of Science and Technology Policy

Staff
Randy Paris
Executive Secretary
Office of Science and Technology Policy

Aeronautics Science and Technology Subcommittee

Chairs
Jaiwon Shin
Associate Administrator
National Aeronautics and Space Administration
Spiro G. Lekoudis
Director, Weapon Systems, Defense Research
and Engineering
Office of the Secretary of Defense

Shelly Yak
Director, William J. Hughes Technical Center
Federal Aviation Administration

Staff
Neal Nijhawan
Executive Secretary
NASA

Member Organizations
Department of Commerce
International Trade Commission
Department of Defense
   Department of the Army
   Department of the Navy
Department of Transportation
   Federal Aviation Administration
   Volpe Center

Department of State
National Aeronautics and Space Administration
National Security Council
Office of Management and Budget
Office of Science and Technology Policy
Office of the U.S. Trade Representative
U.S. International Trade Commission

Alternative Jet Fuel Interagency Working Group

Chairs

Barbara Esker
Deputy Director
Advanced Air Vehicles Program
NASA Aeronautics Research Mission Directorate

Mohan Gupta
Assistant Chief Scientist
Office of Environment and Energy
Federal Aviation Administration

Members

Harry Baumes	United States Department of Agriculture—Office of Energy Policy and New Uses
Dan Birns	United States Department of State, Bureau of Energy Resources—Office of Alternative and Renewable Energy
Nathan Brown	Federal Aviation Administration—Office of Environment and Energy
James (Tim) Edwards	Air Force Research Laboratory—Aerospace Systems Directorate/Turbine Engine Division
Daniel Friend	National Institute of Standards and Technology, Material Measurement Laboratory—Applied Chemicals and Materials Division
William Goldner	United States Department of Agriculture—National Institute of Food and Agriculture—Institute of Bioenergy, Climate, and Environment
Zia Haq	United States Department of Energy—Bioenergy Technologies Office
Aaron Levy	United States Environmental Protection Agency—Office of Transportation and Air Quality
Gregory Rorrer	National Science Foundation—Division of Chemical, Bioengineering, Environmental, and Transport Systems
Bret Strogen	United States Department of Defense—Office of the Under Secretary of Defense for Acquisition, Technology, and Logistics

Table of Contents

Executive Summary

Background

The U.S. civil and military aviation community has historically relied upon energy-dense liquid jet fuel that is derived from petroleum, which is subject to volatile pricing and uncertain supply and harmful to the environment. More recently developed jet fuel alternatives derived from nonpetroleum feedstock can help address those issues and other U.S. energy-related challenges, making development of alternative energy fuels (AJFs) of increased interest in the realm of aviation. Effective Federal research and development can enable AJF development by reducing the costs of producing fuel, facilitating effective evaluation and approval of promising fuel pathways, ensuring that environmental and social benefits accrue from the use of these fuels, reducing technical uncertainty, and promoting private sector investment in production.

To ensure that the United States is most effectively working toward the high-priority research and development (R&D) goals laid out in the 2010 *National Aeronautics Research and Development Plan*, the Aeronautics Science and Technology Subcommittee, under the Committee on Technology of the National Science and Technology Council, established an Alternative Jet Fuel Interagency Working Group (AJF-IWG). The *Federal Alternative Jet Fuels Research and Development Strategy* is the product of the AJF-IWG's work to assess current Federal efforts to address scientific and technical challenges and provide future direction.

Purpose and Scope

The Strategy sets out prioritized Federal R&D goals and objectives to address key scientific and technical challenges that inhibit development, production, and use of economically viable AJFs that would provide energy security and environmental and social benefits relative to conventional fuels, while reducing duplication of effort, enhancing efficiency, and encouraging a coordinated R&D approach among Federal and non-Federal stakeholders. The Strategy complements department and agency R&D policy directives and should guide decisions about R&D program budgets and priorities.

R&D Goals and Objectives

Federal R&D goals are stated below, categorized by their relationship to general components in the path of AJF development.

- Feedstock Development, Production, and Logistics: R&D goals and objectives in this category represent what individual regional supply chains could do to optimize their systems to reduce cost, reduce technology uncertainty and risk, increase yield, and optimize AJF precursors.

  o Increase crop yields (tons/acre), water and nutrient use efficiency, as well as pest and disease resistance, and improve feedstock conversion characteristics

  o Develop sustainable feedstock production systems that require minimal inputs, have a high tolerance for environmental stress, and minimize the risk of adverse environmental impacts (e.g., invasiveness, erosion)

  o Improve harvesting, collection, storage, densification, pretreatment, and transportation of physical biomass to the conversion facility

  o Improve collection, storage, densification, pretreatment, and transportation of municipal solid waste to the conversion facility

- Fuel Conversion and Scale-Up: Fuel conversion and scale-up R&D efforts focus on reducing the cost of production for biochemical, thermochemical, and hybrid conversion processes while increasing the conversion efficiency and volume of fuels produced.

  o Enable discovery, development, enhancement, and scale-up of conversion processes with improved yield, efficiency, and energy requirements that lead to cost-competitive AJF

  o Develop conversion technologies that can produce jet fuel from multiple feedstocks in a distributed manner

- Fuel Testing and Evaluation: Federal R&D efforts in fuel testing and evaluation focus on facilitating the approval of additional AJF pathways by enabling the efficient evaluation of fuel-engine performance and safety through advancement of certification and qualification processes and collection and analysis of data, including those for combustion emissions.

  o Facilitate civil and military approval of additional AJF pathways by enabling efficient evaluation for performance and safety through advancement of certification and qualification processes and collection and analysis of data

  o Improve scientific understanding of how AJF composition impacts gas turbine combustion emissions and operability

- Integrated Challenges: Several key scientific and technical challenges require R&D efforts that either bisect the above components of the AJF development path (i.e., R&D related to feedstock and fuel) or take place outside that path (e.g., during production, deployment, and use). Research in this area requires an interdisciplinary, multi-disciplinary, and multi-faceted approach.

  o Advance understanding of and improve environmental sustainability of AJF production and use

  o Develop and validate a comprehensive systems model to support viable AJF deployment

  o Promote communication as well as scientific and technical R&D best practices for the national enterprise

Other Considerations

If the Strategy R&D goals and objectives are to be accomplished in the time horizon expected, less than 5 years (near term) to more than 10 years (far term), they must be approached with important other considerations in mind.

Non-Technical Challenges

In addition to the scientific and technical challenges associated with AJF development, production, and use, other non-technical challenges are associated with the commercial-scale deployment of AJFs. The benefits of scientific and technical advances can be limited by these challenges, which include volatility in the price of conventional fuels; inadequacies of the production infrastructure; barriers posed by legislation, regulations, and policy; complications of financing structures; uncertainty of investments; and constraints in labor forces and skills. Socio-economic analyses have an important role to play in alleviating such non-technical barriers in this emerging industry, and maximizing the benefit of R&D advances.

Federal Coordination

The AJF-IWG will continue to serve as a focal point for Federal interagency coordination and will work in conjunction with existing formal and informal interagency coordination mechanisms and public-private initiatives to augment coordination efforts.

Public-Private Partnerships

Cooperation between the Federal government and the private sector, including industry, non-governmental organizations, and academia, is crucial to addressing key scientific and technical challenges. Federal agencies should continue to collaborate with non-Federal stakeholders on R&D activities through such mechanisms as stakeholder coalitions, public-private sector initiatives, cost-sharing agreements, and development and demonstration programs. Broad stakeholder engagement is critical to the free flow of information and the development of best practices.

International Coordination

Federal agencies should continue to facilitate international coordination in three primary areas: scientific and technical R&D conducted under multi-lateral and bilateral agreements to mutually share risks, minimize duplication of effort, and benefit from best practices; harmonization efforts to define sustainability criteria to ensure that biofuels achieve desired greenhouse gas reduction goals and do not negatively affect food security and biodiversity; and policy and market-development efforts to ensure a global market for AJFs.

The U.S. Government and industry should continue to cooperate in AJF initiatives that are emerging in diverse countries and participate in AJF activities of the United Nations International Civil Aviation Organization's Committee on Aviation Environmental Protection.

Closing

Continued progress requires focused investments and coordination among Federal departments and agencies, academia, industry, and international partners toward the R&D goals and objectives set out in this Strategy.

Introduction

> Maintaining our leadership in research and development is critical to winning the future and deploying innovative technologies that will create quality jobs and move toward a clean energy economy that reduces our reliance on oil.
>
> *—Blueprint for a Secure Energy Future,* White House, Executive Office of the President (March 30, 2011)

The U.S. civil aviation community and the U.S. military, which are central to the economic well-being and security of the Nation, have demonstrated significant interest in the development and use of alternative jet fuels (AJFs) over the past decade. Jet aviation has historically relied upon energy-dense and relatively inexpensive liquid jet fuels derived from petroleum. As a result of this reliance, U.S. civil aviation and the military face challenges such as price volatility, deleterious environmental impacts, and supply uncertainty. Drop-in[1] AJFs can help address these challenges. The use of these fuels can enhance energy security; expand domestic energy sources; facilitate a diverse, secure, and reliable fuel supply; contribute to price stability; reduce emissions that affect air quality and global climate; generate economic and rural development; and promote social welfare.

An effective Federal research and development (R&D) effort can play an important role in facilitating the development of AJFs. R&D can reduce the costs of producing fuel, facilitate more effective evaluation and approval of promising fuel pathways, ensure that environmental and social benefits accrue from the use of these fuels, and reduce technical uncertainty to a level that will promote significant private sector investment in production capacity.

The development and deployment of alternative fuels (including AJFs), the advancement of biofuels for military and civil transportation, and the promotion of American leadership in renewable energy are national policy priorities.[2] The 2016 Federal Activities Report on the Bioeconomy[3] notes the importance of AJFs as a component of strengthening the U.S. bioeconomy. The 2010 National Aeronautics Research and Development Plan[4] further directs Federal efforts in this area. This plan lays out high-priority aeronautics R&D goals, including the goal to "enable new aviation fuels derived from diverse and domestic resources to improve fuel supply and price stability."[5] In working toward these goals, Federal departments

---

[1] "Drop-in" AJFs are liquid jet fuels that are derived from nonpetroleum feedstock but can replace conventional petroleum-based jet fuel without the need to modify aircraft engines and the fuel distribution infrastructure. In other words, these fuels can be "dropped-in" to the existing system and provide the same level of safety and performance as petroleum jet fuels.

[2] White House, Blueprint for a Secure Energy Future (Washington, DC: Executive Office of the President, March 2011), http://www.whitehouse.gov/sites/default/files/blueprint_secure_energy_future.pdf; White House, National Bioeconomy Blueprint (Washington, DC: Executive Office of the President, April 2012) http://www.whitehouse.gov/sites/default/files/microsites/ostp/national_bioeconomy_blueprint_april_2012.pdf; White House, The President's Climate Action Plan (Washington, DC: Executive Office of the President, June 2013), https://www.whitehouse.gov/sites/default/files/image/president27sclimateactionplan.pdf.

[3] Biomass Research and Development Board (BR&DS), Federal Activities Report on the Bioeconomy, briefing (Washington, DC: Department of Energy, Bioenergy Technologies Office, March 8, 2016), http://biomassboard.gov/pdfs/tac_2016_q1_baumes.pdf.

[4] National Science and Technology Council, National Aeronautics Research and Development Plan (Washington, DC: Executive Office of the President, February 2010), https://www.whitehouse.gov/sites/default/files/microsites/ostp/aero-rdplan-2010.pdf.

[5] Ibid. 39.

and agencies conduct research, develop technologies, build and operate testing infrastructure, and collaborate with non-Federal stakeholders to advance AJF development and its commercial production.

To ensure that the United States is most effectively working toward these goals, the Aeronautics Science and Technology Subcommittee, under the Committee on Technology of the National Science and Technology Council, established an Alternative Jet Fuel Interagency Working Group (AJF-IWG) to assess current Federal efforts, including collaboration with non-Federal stakeholders, and to develop a way forward to address scientific and technical challenges. This Strategy is the product of the AJF-IWG's work.

Purpose and Scope

This report sets out prioritized Federal R&D goals and objectives to address key scientific and technical challenges that inhibit the development, production, and use of economically viable AJFs that would provide environmental and social benefits relative to conventional fuels while enhancing U.S. energy security.

The goals and objectives presented in this Strategy focus Federal R&D efforts to address key scientific and technical challenges while reducing duplication of effort, enhancing efficiency, and encouraging a coordinated R&D approach among Federal and non-Federal stakeholders. This Strategy complements existing department and agency R&D policy directives and should inform Federal R&D program decisions, including budgeting and prioritization.

AJF Development Path

The Development Path shown in Figure 1 represents the process by which an AJF is researched, developed, scaled up, tested, evaluated, and commercialized on a national level. The path begins with an originating raw material or feedstock followed by a conversion process scaled up for production and ends with the fuel product delivered to and consumed by a user. This linear structure echoes the Fuel Readiness Level (FRL) and the Feedstock Readiness Level (FSRL) tools developed by the Commercial Aviation Alternative Fuels Initiative (CAAFI®) to communicate technical development and progress from laboratory to commercial use.[6] Although the Development Path shares some similarities with the concept of the alternative fuel "supply chain," it is different in its emphasis on R&D sectors (e.g., fuel testing) that are not part of the traditional fuel supply chain.

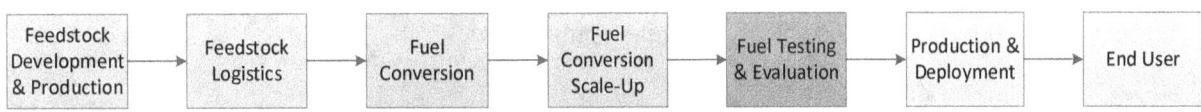

Figure 1. Generalized AJF Development Path

In a supply chain concept, the focus is on an operational production system of an established and approved fuel.[7] In the Development Path for this Strategy, the focus is on technology R&D that will enable AJF approval, production, and commercialization, which would then enable the creation of supply chains.

---

[6] For further information, see CAAFI, "Commercial Aviation Alternative Fuels Initiative: Fuel Readiness Tools" (2016), http://www.caafi.org/information/fuelreadinesstools.html.

[7] The U.S. Department of Agriculture (USDA) defines "supply chain" as feedstock crop development, production, and logistics (harvest, densification/diminution, storage, pretreatment, and transportation to biorefinery conversion platform), conversion, marketing and distribution, and end use.

The components of the Development Path are described as follows:

- Feedstock Development and Production: Identification, assessment, and improvement of potential AJF raw materials (feedstocks) including bio-feedstock and fossil-fuel feedstock.

- Feedstock Logistics: Development of efficient means of collecting, harvesting, and moving feedstock; preprocessing feedstock; and other activities necessary to prepare feedstock before conversion to fuel.

- Fuel Conversion: Development and improvement of a feedstock-to-fuel conversion process.

- Fuel Conversion Scale-Up: Advancement of a fuel conversion process to a higher volume of production beyond laboratory and pilot-scale levels to support commercialization.

- Fuel Testing and Evaluation: Fuel testing and evaluation is usually iterative, requiring increasing volumes, and may occur concurrently with fuel conversion development and scale-up.

  o Fuel Performance Testing and Evaluation: Determination of fuel characteristics and engine system performance when using the AJF, including testing whether the fuel meets qualification and certification requirements.

  o Emissions Characterization Testing: Analysis to understand fuel-use impacts on air quality and global climate.

- Production and Deployment: Initial commercial-scale availability of fuels for end-use purposes.

- End User: Purchase and use of the AJF by aviation fuel users such as commercial airlines and the U.S. Government.

While the Development Path is a useful linear model, it does not completely capture the complexity of the AJF R&D enterprise. In practice, each component of the Development Path is connected to and influences the other components. For example, feedstock R&D is influenced, in part, by the state of science and the technologies for moving and converting feedstock to fuel and other products. The state of science and technologies in feedstock conversion, in turn, affect the quality and quantity of feedstock available for fuels conversion and the R&D to scale-up for commercial production. In addition, complex issues such as environmental sustainability and techno-economic considerations of AJF production and use encompass the entire Development Path as an integrated system. To ensure that the collective R&D enterprise is effective and efficient, AJF R&D efforts must be informed by an awareness of this complexity and the interrelated nature of the links in the Development Path.

R&D Goals and Objectives

The AJF Development Path shown in Figure 1 describes the key areas in which these R&D goals and objectives can positively impact and accelerate AJF development. Efforts to develop and deploy AJFs have evolved significantly over the last decade, particularly through federally funded activities, regional group efforts, and international forums. The R&D goals and objectives of this Strategy leverage these prior efforts, building on the progress made in advancing the state of knowledge and insights on scientific and technical challenges.

R&D goals and objectives for this Strategy are categorized into the following areas, as depicted in Figure 2:

- Feedstock Development, Production, and Logistics

- Fuel Conversion and Scale-Up

- Fuel Testing and Evaluation

- Integrated Challenges

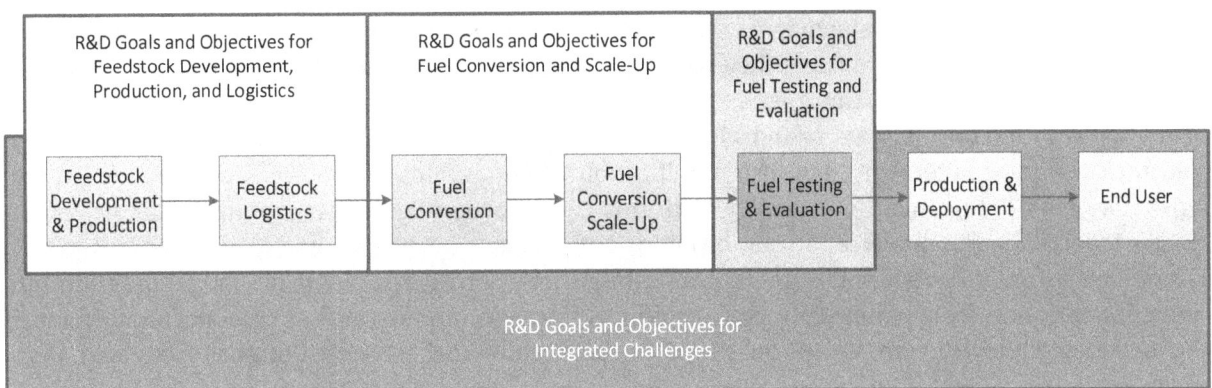

Figure 2. Overview of R&D Goals and Objectives Alignment to the Generalized AJF Development Path

Each set of goals and objectives is intended to guide R&D efforts in addressing key scientific and technical challenges. Based on the availability of resources and annual appropriations, these R&D goals and objectives are projected to be completed over near-term (<5 years), mid-term (5–10 years), and far-term (>10 years) time horizons. Appendix 3 contains a table of these R&D goals and objectives and includes an identification of departments and agencies with an R&D mission that aligns with a particular goal or objective.

Feedstock Development, Production, and Logistics

Goal 1: Increase crop yields (tons/acre), water and nutrient use efficiency, as well as pest and disease resistance, and improve feedstock conversion characteristics

Goal 2: Develop sustainable feedstock production systems that require minimal inputs, have a high tolerance for environmental stress, and minimize the risk of adverse environmental impacts (e.g., invasiveness, erosion)

Goal 3: Improve harvesting, collection, storage, densification, pretreatment, and transportation of physical biomass to the conversion facility

Goal 4: Improve collection, storage, densification, pretreatment, and transportation of municipal solid waste to the conversion facility

R&D goals and objectives in this category represent what individual regional supply chains could do to optimize their systems to reduce cost, reduce technology uncertainty and risk, increase yield, and optimize AJF precursors. The objectives include a two-tier approach, facilitating both discrete R&D within the feedstock supply chain component and integrated R&D between the feedstock and conversion components. Consequently, by working back from targets and factoring unique resources and capacities, these goals and objectives provide an integrated R&D approach across regional feedstock supply chains linked to one or more commercial product conversion systems. The optimization of individual regional supply chain systems will depend, in part, on the region, feedstock, and conversion process. All types of feedstock are included in this Strategy, provided that the feedstock has the potential to be converted to drop-in AJF while having environmental and societal advantages over petroleum-based jet fuel. For MSW, the primary challenges are associated with logistics (materials handling) and not with feedstock supply per se.

Setting up a feedstock supply system requires integration and iteration. The near-, mid-, and long-term goals outline this approach. The genetic improvement of feedstocks, whether agronomic or woody, requires an integrated understanding of the entire feedstock system to identify genetic improvement targets. As feedstocks are developed, they need to be evaluated and selected genotypes moved forward to be optimized in production, logistics, and conversion systems. Feedback loops from feedstock evaluation with the downstream systems inform not only the genetic targets for the next generation of feedstock genotypes (iterative), but also modifications/optimizations of the feedstock production and logistic systems (integrative). Conversion platforms can also be modified to optimize AJF yield from specific feedstocks. This modification may be important, for example, when feedstock characteristics such as inorganic ash content reduce the efficiency of catalysts or when feedstock organic compounds interfere with enzymatic activity in bio-catalytic conversion platforms. As new system components are developed and tested, technology transfer specialists will require additional training alongside new specialists, engineers, foresters, and agronomists.

AJF production in sufficient quantities and at a competitive price is the general goal, with impetus coming from the commercial aviation industry and the military. A regional focus may help identify specific production opportunities and define cost competitiveness. Planning for regional AJF systems must integrate feedstock supply system elements with AJF production and end-use by identifying the following:

- Fuel use locations (e.g., Atlanta/Hartsfield International Airport, military air stations in Hawaii);

- Region(s) likely to benefit most from producing or consuming AJFs (due to the economic, social, and environmental characteristics of AJF);

- Regionally appropriate feedstocks (e.g., southern pines, hybrid poplar, energy cane, perennial grasses, MSW, forest/mill residuals, oil crops, and waste greases);

- Extant/emerging conversion platforms that could use these types of feedstock;

- Industry/community interest in siting a biorefinery in the region; and

- Alternative uses/products potentially supported by the feedstock supply chain (i.e., potential for synergistic economics/competition).

Individual aspects of the supply chain may be addressed to derive incremental improvement. However, because feedstock/conversion supplies systems require feedback loops, designing and implementing integrated approaches may be the quickest way to stand up a new AJF industry.

Fuel Conversion and Scale-Up

Goal 1: Enable discovery, development, enhancement, and scale-up of conversion processes with improved yield, efficiency, and energy requirements that lead to cost-competitive AJF

Goal 2: Develop conversion technologies that can produce jet fuel from multiple feedstocks in a distributed manner

Fuel conversion and scale-up R&D efforts focus on reducing the cost of production for biochemical, thermochemical, and hybrid conversion processes while increasing the conversion efficiency and volume of fuels produced. Additional challenges are associated with the scale-up of conversion technologies from the laboratory to commercial scales. The specific issues that need development include process integration, the challenge and complexity of demonstration and pioneer plants coming on line to produce volumes of fuels to enable jet fuel testing, and the production of jet fuel from these facilities at cost-competitive levels.

Technologies are being developed at various scales (pilot, demonstration, and pre-commercial) to convert biomass feedstocks into jet fuel, other fuels, and chemicals. Conversion technologies that are relatively mature include (1) hydro-treatment and upgrading of waste oils or plant-based oils to jet fuel and (2) gasification of biomass or MSW into a synthesis gas followed by Fischer-Tropsch conversion of the synthesis gas into jet fuel. However, R&D is needed even in these relatively mature technologies. For example, the price of fuels from hydro-treatment of oils is dominated by the cost of the feedstock, which can account for 75 to 80% of the cost of the finished fuel. The availability of waste-based oil feedstocks is limited and imposes an upper bound on potential production volumes. R&D could focus on new feedstocks that can be available at low costs to make the finished fuel cost competitive. The gasification/Fischer-Tropsch technologies have high capital costs and require large facilities to achieve economies of scale. R&D could enable modular, small-scale reactors that can convert bio-derived synthesis gas into jet fuel. One promising pathway being explored is producing a bio-crude from biomass or non-fossil feedstocks and co-processing the bio-crude and fossil-based crude oil in existing refineries. The refinery would continue to produce a mix of products, including jet fuel, diesel, gasoline, and petrochemicals.

Conversion technologies in the mid-term time horizon include alcohol-to-jet (ATJ), additional biochemical/catalytic conversion of sugars to hydrocarbons, and pyrolysis. In the ATJ process, ethanol from existing biorefineries can be catalytically converted into jet fuel. The advantage of this approach is that it can leverage existing biorefinery infrastructure, such as wet or dry corn mills, and emerging cellulosic biorefineries. One area of active R&D is the co-production of high-value bio-based chemicals, fertilizers, and soil supplements as a means of reducing the cost of jet fuels. For biochemical processes, specific areas of R&D include new and improved enzymes, microorganisms, and processes that can combine pre-treatment and conversion steps in one reactor. For thermochemical processes, specific areas of R&D include development of durable catalysts that are easy to regenerate, hot gas clean-up technologies, and reliable feeding of biomass into pressurized, high-temperature gasifiers. A near-term opportunity could be the production of bio-crude that is suitable for integration into an existing petroleum refinery.

Conversion technologies in the long-term time horizon include conversion of waste carbon dioxide (CO_2) into ethanol followed by ATJ conversion, processes involving algal and other microbial feedstocks, and algae or other microbes capable of producing hydrocarbons. R&D challenges that need to be addressed include new strains and culture-management approaches; high-productivity cultivation systems that maximize yields while minimizing water, land, and nutrients; and low-cost, high-throughput drying and harvesting technologies that can be easily integrated with cultivation systems.

Fuel Testing and Evaluation

Goal 1: Facilitate civil and military approval of additional AJF pathways by enabling efficient evaluation for performance and safety through advancement of certification and qualification processes and collection and analysis of data

Goal 2: Improve scientific understanding of how AJF composition impacts gas turbine combustion emissions and operability

Testing and evaluation must be completed successfully for a fuel to be approved for use by commercial aviation and the military. Federal R&D efforts in fuel testing and evaluation focus on facilitating the approval of additional AJF pathways by enabling the efficient evaluation of fuel-engine performance and safety through advancement of certification and qualification processes and collection and analysis of data, including those for combustion emissions. Goals to achieve this purpose reflect an R&D approach focused on reducing the cost, time, and uncertainty of the fuel testing and evaluation process to enable efficient and timely certification, qualification, and acceptance of candidate AJFs.

A streamlined fuels testing and evaluation process through the development of generic rig combustion and numerical modeling capability is urgently needed. This generic fuels testing and evaluation methodology would produce engine operability and performance data comparable to data from actual testing of several engines from different manufacturers. It will help reduce the resource requirement significantly and will allow more efficient and faster certification and qualification in the future. It will also promote opportunities for exploration of additional fuel candidates for commercial use.

Research on the effects of fuel composition on combustion emissions from gas turbines through measurements—at ground and at cruise altitude—must be continued to examine the potential emissions benefits that new fuel candidates may provide. Emissions measurements are resource intensive; therefore, model-based predictive capability of combustion emissions must be developed and validated against the measurement data.

Integrated Challenges

Goal 1: Advance understanding of and improve environmental sustainability of AJF production and use

Goal 2: Develop and validate a comprehensive systems model to support viable AJF deployment

Goal 3: Promote communication as well as scientific and technical R&D best practices for the national enterprise

Integrated challenges are key scientific and technical challenges that require R&D efforts across multiple sectors of the Development Path or are outside the scope of the linear Development Path model. The research in this area requires an interdisciplinary, multi-disciplinary, and multi-faceted approach.

Environmental sustainability is a commonly accepted societal good but remains difficult to measure, model, and predict even if subcomponents can be quantified (e.g., health impacts of sulfur oxides (SO_x), nitrogen oxides (NO_x), particulate matter). The environmental performance of a product or service is perhaps more quantifiable and less dependent on societal dynamics, but, in this document, we use the terms "environmental sustainability" and "environmental performance" interchangeably. The overall environmental impacts of AJFs can only be understood by developing methods and measurements that can be applied consistently across the entire enterprise. An identified challenge is the need for an integrated analysis of environmental sustainability as applied to AJFs.

Adverse environmental impacts tend to be highest in the areas of feedstock production, fuel production, and end use. Establishing and validating protocols that can be used for measuring environmental impacts in a common and consistent manner throughout the supply chain is an important goal. The scientific bases for the measurement of each indicator of environmental sustainability must be technically sound. Key outcomes of this research would be the development of common definitions, protocols, and uncertainty methodologies for assessing and reporting each aspect of environmental sustainability.

A systems model of the full AJF life cycle considers feedback and interactions, moving beyond the simplified linear representation. Some techno-economic and regional models have been developed, but a full systems model for the AJF life cycle considers integrated environmental, economic, and social aspects. For example, a useful systems model must estimate the effects of the AJF industry on rural job creation and the price and supply of national and global food commodities in the scenarios that can be envisaged. Building a comprehensive systems model requires the identification and quantification of elements and interactions, the development of analytical tools, and the implementation of a unified framework from its constituent elements. Techno-economic models can be used to project whether certain feedstocks and conversion pathways appear to be commercially viable and can guide future R&D investments to lower fuel production cost.

A successful model will contribute toward plug-and-play engineering applications, with validated analyses supporting decision making among options related to cost and sustainability as related to region, pathway, logistics, feedstock, and so forth.

Although a number of resources exist for information on AJFs, there is no centralized repository or portal. To better link researchers and other stakeholders involved in AJF development and to better disseminate best practices, a publically accessible repository of information would be beneficial and consistent with the Federal government's policy of increasing access to the results of federally funded scientific research.

Non-Technical Challenges

The commercial-scale deployment of AJFs also faces challenges that are not specifically scientific or technical in nature. The evolving demand for conventional and alternative fuels is influenced, in part, by price volatility of conventional jet fuel; production infrastructure barriers; legislative, regulatory, and policy barriers; complicated financing structures; investment uncertainty; and labor force and skill constraints. It is important to recognize that the benefits of scientific and technical R&D advances could be limited by such non-technical challenges. By recognizing the broader context for this emerging industry and the roles of non-technical research (e.g., socio-economic analyses), the impact of the scientific and technical R&D can be maximized. The developing industry may benefit if additional progress is made in understanding the relationship between feedstock and fuel prices and price volatility; projecting future production and demand for conventional and AJFs; evaluating the effectiveness and political economy of policy options; identifying legislative, regulatory, public perception, and infrastructure barriers; optimizing financing structures to mitigate investment and other economic uncertainty; and managing labor force and skill constraints. Complementarily, scientific and technical R&D can inform policy decisions and can help reduce technical risks and potentially mitigate economic and financial risks.

Federal Coordination

To enable continued coordination of Federal AJF R&D efforts, the AJF-IWG will continue to be chartered under the auspices of the National Science and Technology Council's Aeronautics Science and Technology Subcommittee. This body will continue to include representatives from appropriate Federal agencies and will serve as a focal point for Federal interagency coordination and will work in conjunction with existing formal and informal interagency coordination mechanisms and public-private initiatives to augment those efforts. The AJF-IWG will ensure that duplication is avoided since it will have a lead role in the implementation of this Strategy, including informing program, budgeting, and prioritization decisions; coordinating activities; identifying outcomes (e.g., joint competitive solicitations, joint review of proposals); performing progress assessment; engaging stakeholders; and making recommendations to agencies toward meeting evolving R&D needs.

Public-Private Partnerships

Cooperation between the Federal government and the private sector, including industry, non-governmental organizations (NGOs), and academia, is crucial to addressing key scientific and technical challenges. Federal agencies have collaborated with and should continue to collaborate with non-Federal stakeholders on R&D activities through a variety of mechanisms, including stakeholder coalitions, public-private sector initiatives, cost-sharing agreements, and development and demonstration programs. Broad stakeholder engagement, such as that represented by CAAFI® and "Farm to Fly 2.0,"[8] are critical to enabling coordinated efforts of the private and public sectors to develop AJFs. These mechanisms provide a means of the free flow of information and the development of best practices, and they provide critical forum for establishing common direction in R&D efforts.

International Coordination

Aviation is a global industry by nature and, as such, technology and support infrastructure (e.g., aviation fuels) must transcend national boundaries to provide a seamless transportation system. Therefore, Federal agencies have facilitated and should continue to facilitate international coordination in three primary areas: scientific and technical R&D conducted under multi-lateral and bilateral agreements to mutually share risks, minimize duplication of effort, and benefit from best practices; harmonization efforts to define sustainability criteria to ensure that biofuels achieve desired greenhouse gas (GHG) reduction goals and do not negatively affect food security and biodiversity; and policy and market-development efforts to ensure a global market for AJFs.

AJF initiatives have emerged in countries as diverse as Australia, Brazil, Canada, China, Finland, Germany, Iceland, Indonesia, the Netherlands, Norway, Sweden, Spain, and the United Arab Emirates, among others. The U.S. Government and industry have established cooperation and coordination with many of these efforts to leverage common science and technology interests and investments and mutually advance technical knowledge. U.S. agencies also actively participate on AJF activities of the United Nations International Civil Aviation Organization (ICAO) Committee on Aviation Environmental Protection (CAEP). Additional cooperation and coordination takes place between U.S. public/private initiatives, such as CAAFI®, and efforts of the Australian Initiative for Sustainable Aviation Fuels (AISAF), the Brazilian Biofuels Platform (BBP), the Aviation Initiative for Renewable Energy in Germany (Aireg), Indonesia's Aviation Biofuels and Renewable Energy Task Force, and Spain's Bioquereseno initiative.

Conclusions

AJFs can help enhance energy security; expand domestic energy sources; facilitate a diverse, secure, and reliable fuel supply; contribute to price and supply stability; reduce emissions that affect air quality and global climate; generate economic and rural development; and promote social welfare. Federal R&D plays an important role in facilitating the development of AJFs. This Strategy focuses Federal R&D efforts to address key scientific and technical challenges while reducing duplication of effort, enhancing efficiency, and encouraging a coordinated R&D approach among Federal and non-Federal stakeholders. Continued progress requires focused investments and coordination among Federal departments and agencies, academia, industry, and international partners toward the R&D goals and objectives set out in this Strategy.

---

[8] See Appendix 2 for information on CAAFI and Farm to Fly 2.0.

Appendix 1.
Agency-Specific Contributions to Research and Development (R&D) of Alternative Jet Fuels (AJFs)

Figure A-1 identifies Federal agencies that have an R&D mission that aligns with a particular area of the AJF Development Path. When multiple agencies are indicated in the same area, it means that the agencies are working in complementary manner with distinct aspects of the research area. This is facilitated by regular coordinated or joint funding of activity to avoid duplication of effort.

	Feedstock Development & Production	Feedstock Logistics	Fuel Conversion	Fuel Conversion Scale-Up	Fuel Testing & Evaluation	Integrated Challenges
DOC	X					X
DoD				X	X	
DOE	X	X	X			X
DOT					X	X
EPA						X
NASA					X	
NSF	X	X	X			
USDA	X	X	X			X

Figure A-1. Agency-Specific Contributions to AJF Development Path

Department of Commerce (DOC)

DOC, primarily through the work of the National Oceanic and Atmospheric Administration (NOAA), the National Institute of Standards and Technology (NIST), and the International Trade Administration (ITA), supports R&D across the AJF Development Path to promote job creation, economic growth, and sustainable development associated with the AJF industry. This support is accomplished by advancing measurements and standards, promoting technologies in the global arena, and providing atmospheric and terrestrial measurements in support of decisions relating to the role of aircraft emission sources and land-use changes. DOC research has focused on alternative fuel properties and aviation emissions and their effects on the atmosphere. Other activities have involved deoxyribonucleic acid (DNA) and metabolomics for optimizing the next generation of crops, optimizing and characterizing catalysts and catalysis processes, and information and algorithms for simulating conversion processes. DOC will continue its Earth-observing mission in support of biomass-based jet fuels; address the identified needs for measurements, models, reference materials, and property databases on biofuels; and support innovation and efficiencies in the AJF sector throughout the Development Path.

Department of Defense (DOD)

DOD supports the testing and approval of AJF to enable the broadest set of available energy options and supports national goals for energy security and assured operations. It conducts research, development, testing, and evaluation for aircraft propulsion, airframe subsystems, and ground infrastructure to enable alternative fuel use in DOD platforms; develops knowledge-based certification methods to reduce qualification and certification time and cost; conducts R&D to identify opportunities that will enhance the process of qualifying soon-to-be commercially available alternative fuels for use in military platforms; and evaluates the airworthiness of AJFs for use in military platforms. DOD also performs testing relevant only to military uses, which include evaluating performance of AJFs in ground-based diesel engines and afterburning fighter engines. The Departments of the Army, Navy, and Air Force will continue to evaluate pathways for potential incorporation of AJF into military and commercial jet fuel specifications. DOD will continue to support the Defense Production Act (DPA) Title III Advanced Drop-in Biofuel Production Project, through which three drop-in fuel producers are receiving $70-million awards that were jointly funded by the DOD and DOE. Furthermore, the Defense Logistics Agency (DLA), on behalf of the Navy (and in partnership with the United States Department of Agriculture (USDA) through the Farm-to-Fleet initiative), recently began to solicit for deliveries of JP-5 jet fuel that include up to 50% bio-based AJF starting in 2015. This bio-based AJF procurement initiative is expected to continue in the coming years

Under DOD Directive (DODD) 4180.01, *DOD Energy Policy*, it is stated that DOD will "diversify and expand energy supplies and sources, including renewable energy sources and alternative fuels."[9] The Navy has set a goal of alternative sources used for 50% of total Navy energy consumption afloat (which includes jet fuel) by 2020 (approximately 336 million gallons of alternative fuels annually, including marine and jet fuels). The Air Force has set a goal to increase the use of cost-competitive drop-in AJF blends for non-contingency operations to 50% of total consumption by 2025 (likely to exceed 1 billion gallons annually).

Department of Energy (DOE)

DOE supports basic and applied R&D, analysis, demonstration, and deployment efforts for next-generation AJFs. These efforts in biofuels are aligned with DOE's priorities of science and discovery, clean energy, economic prosperity, and mitigation of climate change. DOE's programs in the Office of Energy Efficiency and Renewable Energy, the Office of Science, and the Advanced Research Projects Agency-Energy (ARPA-E), as well as DOE's Loan Guarantee Program, have made investments in next-generation biofuels. This work and other efforts help form a foundation for DOE's continued support and development of commercially viable and sustainable AJF conversion technologies. Specific DOE activities in applied areas include applied R&D to reduce the cost and risk associated with innovative pathways to AJF, including developing catalysts and biocatalysts and upgrading biomass-derived intermediates; cost-shared demonstration of these technologies at pilot and demonstration scale; collaborating on certification, testing, and approval processes for AJF from sustainable feedstocks; assessing feedstock resources; and conducting techno-economic analysis of innovative alternative conversion technologies. DOE also supports basic research through the Bioenergy Research Centers, Energy Frontier Research Centers, targeted solicitations, and ongoing programs that advance fundamental knowledge and technological innovation for the development of AJF. Such efforts address fundamental science issues important for feedstock production, biomass quality, catalyst development, and feedstock conversion. Examples of basic research activities include structural and mechanistic studies that can inform the design of highly selective and efficient catalysts; analysis of plant cell wall biosynthesis and structure to guide

---

[9] Department of Defense, "DoD Energy Policy," Department of Defense Directive (DODD) 4180.01 (Washington, DC: USD (AT&L), April 16, 2014), 1, http://www.dtic.mil/whs/directives/corres/pdf/418001_2014.pdf.

feedstock conversion strategies; development of chemical, enzymatic, and pyrolytic approaches for biomass conversion; studies of photosynthetic efficiency and regulation to enhance biomass production; and understanding of plant lipid biosynthesis for increased oil production. DOE is also a signatory agency in the "Farm to Fly 2.0" agreement with the Federal Aviation Administration (FAA) and USDA.

Department of Transportation (DOT)

Within DOT, the FAA leads R&D efforts to address barriers and support the adoption, production, and end use of AJFs by the civil jet aviation community to enable environmentally unconstrained growth of commercial aviation. Agency efforts include AJF performance testing to inform ASTM International approval; emissions testing; analyses of environmental impact and economic and supply potential; and support to facilitate coordination between the public, private, and research sectors. Current activities include coordination of the aviation industry, government, academic, and energy industry stakeholders through CAAFI®; support for AJF testing for ASTM International qualification by industry via the Continuous Lower Energy, Emissions, and Noise (CLEEN) technology development program; and studies related to aircraft emissions, environmental impacts, and supply chain development conducted by the Aviation Sustainability Center (ASCENT), an FAA-sponsored university-based center of excellence on AJF and Environment, to address knowledge gaps and inform decision-making. The DOT Volpe National Transportation Systems Center provides valuable technical support, which builds on its long history of evaluating the national transportation system to conduct environmental and economic analyses of AJFs. DOT will continue to support coordination of efforts via CAAFI® and to research ways to enable AJFs, including analysis of the opportunities and barriers to establishing AJF supply chains in multiple regions of the United States and environmental and techno-economic analysis of multiple AJF pathways in relation to conventional jet fuels. The FAA has set an aspirational goal for the U.S. aviation industry to use 1 billion gallons of AJFs annually by 2018 (about 5% of expected use).

Environmental Protection Agency (EPA)

EPA is developing and implementing the Renewable Fuel Standard (RFS) program to ensure that transportation fuel sold in the United States contains a minimum volume of renewable fuel. Under the RFS program, jet fuel derived from renewable biomass can qualify for tradable credits known as Renewable Identification Numbers (RINs) if it meets certain criteria. In the March 2010 RFS final rule (75 Federal Register 14670),[10] EPA determined that a number of jet fuel pathways qualify as advanced biofuel or cellulosic biofuel based on their life-cycle reductions in GHG emissions compared to baseline fuel and the feedstocks used to produce them. In addition, through regulations issued in March 2013 (78 Federal Register 14190),[11] EPA clarified that some renewable diesel processes that had been previously evaluated included jet fuel and also approved additional jet fuel pathways. In the March 2010 RFS rule, EPA also created a petition process whereby parties can petition the agency to evaluate the life-cycle GHG emissions of additional jet fuel pathways. Through this petition process, the EPA will continue to use the best available science to evaluate new jet fuel pathways under the RFS program.

---

[10] Regulation of Fuels and Fuel Additives: Changes to Renewable Fuel Standard Program; Final Rule, 75 Fed. Reg. 14670 (March 26, 2010), https://www.gpo.gov/fdsys/pkg/FR-2010-03-26/pdf/2010-3851.pdf.

[11] Regulation of Fuels and Fuel Additives: Identification of Additional Qualifying Renewable Fuel Pathways under the Renewable Fuel Standard Program, 78 Fed. Reg. 14190 (March 5, 2013), https://www.gpo.gov/fdsys/pkg/FR-2013-03-05/pdf/2013-04929.pdf.

National Aeronautics and Space Administration (NASA)

Aeronautics research at NASA includes developing knowledge, technologies, tools, and innovative concepts to enable safe, new generations of civil aircraft that are more energy efficient and have a smaller environmental footprint. Advanced, environmentally friendly aircraft will be needed in the future to support domestic and international air transportation demands while protecting and preserving the environment. To this end, NASA has been working with other government agencies and organizations to characterize the environmental benefits of "drop-in" AJFs by measuring the emissions of these fuels from actual aircraft engines on the ground and in-flight. Data resulting from these tests are being used as input to air quality and climate change models and are informing the development of aviation emission standards. NASA research also includes continuing to advance the physics-based understanding of combustion and combustion processes as a foundation for ensuring that future aircraft systems can take full advantage of the environmental benefits offered by AJFs. NASA will also continue to investigate potential technology solutions for the more distant future. These solutions may dramatically reduce fuel usage and leverage more electric power.

National Science Foundation (NSF)

NSF supports basic research in all non-medical fields of fundamental science and engineering. NSF's clean energy investments include research related to sustainability science and engineering (e.g., the conversion, storage, and distribution of diverse power sources and the science and engineering of energy materials, energy use, and energy efficiency). As part of these clean energy activities, NSF supports basic research to advance the innovation and fundamental understanding needed for the development of advanced biofuels, including feedstock production, feedstock logistics, and feedstock conversion. NSF supports fundamental research to enable biomass feedstock production through ongoing programs that support the plant sciences, and targeted solicitations, most notably the Plant Genome Research Program (PGRP). Research activities supported by the PGRP focus on fundamental questions in plant sciences and the development of sequence resources in crop plants. NSF also supports research to address fundamental scientific and technological issues related to biomass feedstock logistics, including environmental and sustainability challenges of plant biomass harvesting; systems-level analysis and life-cycle assessment of integrated bio-refinery operations; and innovative approaches for algae harvesting and dewatering for algal biofuel production processes. NSF supports scientific and engineering research to drive innovation and fundamental understanding of biological and chemical routes for the conversion of renewable feedstocks into advanced biofuels. Relevant examples include new catalysts for the conversion of biomass constituents into hydrocarbons and for the direct reduction of CO_2 and water into energy-dense molecules; advanced chemical reactor processes for thermochemical or thermocatalytic breakdown and deoxygenation of biomass into hydrocarbons; genetic engineering and synthetic biology of microorganisms for advanced biofuel production pathways from renewable feedstocks; and bioprocess engineering of native and genetically engineered microorganisms (bacteria, yeast, and algae) for production of alcohols, lipid-based biofuels, or advanced hydrocarbon biofuels from biomass-derived sugars or from sunlight and CO_2.

United States Department of Agriculture (USDA)

USDA enables the production and use of dependable supplies of dedicated feedstocks for the sustainable production of aviation biofuels through the creation of new scientific knowledge and technological innovation and program and policy support. The Agricultural Research Service (ARS) and the Forest Service lead internal efforts through the USDA Regional Biomass Research Centers that support the development of region-based strategies for the sustainable production of dedicated feedstocks and their genetic improvement. The ARS provides scientific leadership that directly advances technologies to enable AJF conversion from non-food feedstocks. The National Institute of Food and Agriculture (NIFA) provides leadership and funding for extramural research, development, demonstration, education, and outreach activities through competitive programs (e.g., the Agriculture and Food Research Initiative (AFRI) Sustainable Bioenergy Challenge, the Biomass Research and Biomass Research and Development Initiative (BRDI), the USDA/DOE Plant Feedstock Genomics for Bioenergy Program, and the Small Business Innovation Research (SBIR) Program and through non-competitive programs that target workforce and human capital development). Other USDA programs support commercial deployment of feedstock production, logistic, and conversion systems; availability of sustainable feedstock supplies; and research on indirect land-use and GHG emissions. USDA will continue extensive intramural and extramural investment in complete feedstock supply systems, demonstration pilot plants, biorefineries, and distribution infrastructure across the biofuels spectrum. Foundational, applied, and pre-commercialization R&D, demonstration, sustainability (economic, environmental, and social) analyses, work force development, technology transfer, policy information and analysis, and risk reduction and management are among the activities that USDA will continue to support in the future.

Appendix 2.
Multi-Agency Activities that Contribute to Research and Development (R&D) of Alternative Jet Fuels (AJFs)

The agencies also support joint initiatives and inter-agency and government-industry coordination efforts that advance AJF development and its future commercial production. For example, the Federal Aviation Administration (FAA) and the Department of Defense (DOD) support research to determine the technical feasibility of using new AJFs in aircraft with the existing infrastructure. While many of the tests that DOD performs resemble those required by ASTM International, DOD also evaluates AJF blends for use in aircraft that perform under conditions atypical for commercial aircraft (e.g., afterburning fighter engines) in addition to testing these fuels for use in its ground fleet of (primarily) diesel engines. Also, the U. S Department of Agriculture (USDA), the Department of Energy (DOE), and the Department of the Navy signed a Memorandum of Understanding (MOU) in 2011 in which they coordinate activities to support the construction, retrofit, and operation of a selected set of domestic commercial-scale production facilities to produce advanced drop-in biofuels, including AJFs that meet military specifications. Three facilities have been selected and are proceeding with the design phase. Details of these and other example initiatives are given below.

Biomass Research and Development Initiative (BRDI)—The Biomass Research and Development Board[12] is an interagency collaborative composed of senior decision-makers from Federal agencies and the White House, co-chaired by USDA and DOE. In addition to USDA and DOE, the Board includes members from the Department of Transportation (DOT), the Environmental Protection Agency (EPA), the National Science Foundation (NSF), the Department of Interior (DOI), DOD, and the Office of Science and Technology Policy (OSTP). It was established by the Biomass Research and Development Act of 2000,[13] later amended by Section 9001 of the Food Conservation and Energy Act of 2008 (FCEA),[14] and most recently reauthorized in the Agricultural Act of 2014.[15] The Board annually implements the Biomass Research and Development Initiative (BRDI),[16] which consists of grants made available through FCEA Section 9008,[17] and other programs. The Technical Advisory Committee[18] (Committee or TAC) is an independent body that provides input to agencies regarding the technical focus and direction of the Initiative.

Advanced Drop-In Biofuels Production—In August 2011 the USDA, DOE, and Navy proposed to invest directly with industry in infrastructure to produce advanced drop-in aviation and marine biofuels to power military and commercial transportation. The effort consists of:

---

[12] Biomass Research and Development, "Board," last updated 03/16/2015, http://www.biomassboard.gov/board/board.html.

[13] Office of Energy Efficiency & Renewable Energy, "Biomass Research and Development Act of 2000," http://energy.gov/eere/bioenergy/downloads/biomass-research-and-development-act-2000.

[14] Food Conservation and Energy Act of 2008, Pub. L. 110-246, 121 Stat. 2064 (2008), https://www.gpo.gov/fdsys/pkg/PLAW-110publ246/pdf/PLAW-110publ246.pdf.

[15] Agricultural Act of 2014, H.R. 2642, 113th Cong. (2014), https://www.gpo.gov/fdsys/pkg/BILLS-113hr2642enr/pdf/BILLS-113hr2642enr.pdf.

[16] Biomass Research and Development, "Initiative," last updated 12/09/2010, http://www.biomassboard.gov/initiative/initiative.html.

[17] Food Conservation and Energy Act of 2008, Pub. L. 110-246, 121 Stat. 2079 (2008), https://www.gpo.gov/fdsys/pkg/PLAW-110publ246/pdf/PLAW-110publ246.pdf.

[18] Biomass Research and Development, "Advisory Committee," last updated 03/16/2015, http://www.biomassboard.gov/committee/committee.html.

- *Defense Production Act (DPA)*—The Department of the Navy, USDA, and the DOE are collaborating with industry to co-fund the construction of three fuel production facilities through the Defense Production Act of 1950 (DPA), as amended (50 U.S.C. App. § 2061 et seq.).[19] Title III of the DPA enables this funding mechanism, thereby identifying the development of alternative fuel production capability to be of strategic importance to the security of the United States. Funding for the three facilities that will produce 100 million gallons (combined) of diesel and jet fuel was announced in September 2014.

- *Farm to Fleet*—In summer 2014, DOD published a bulk fuel solicitation that, for the first time, encourages biofuel blends of 10% or greater in F-76 and JP-5 fuels. USDA is supporting this DOD solicitation by offering Commodity Credit Corporation (CCC) funds to cover the price differential between alternative and conventional jet (or diesel) fuels.

Farm to Fly 2.0 Agreement—In April 2013, USDA and DOT signed a "Farm to Fly 2.0" agreement with the aviation industry represented by Airlines for America (A4A), Aerospace Industries Association (AIA), Airports Council International-North America (ACI-NA), General Aviation Manufacturers Association (GAMA), and the National Business Aviation Association (NBAA) to work together toward achieving the aspirational goal of having 1 billion gallons of AJF in use by U.S. civil and military aviation by 2018. In July 2014, DOE joined as a signatory to the agreement. CAAFI® provides facilitation support. The initiative focuses government and aviation industry resources toward jet fuel supply chain development, including carrying out feasibility studies for fuel production in various U.S. states.

Feedstock Readiness Level Tool—The USDA and the FAA signed an MOU on Aviation Biofuel Development in October 2010 to create a framework of cooperation to leverage expertise of the two agencies to develop AJF production. It was a tri-partite agreement between FAA's Office of Environment and Energy, the USDA's Agricultural Research Service (ARS) and the USDA Office of Energy Policy and New Uses (OEPNU). Under the partnership, the three offices developed the Feedstock Readiness Level (FSRL) Tool to assess availability of different kinds of feedstocks that will be needed by biorefineries to produce renewable jet fuels.

Commercial Aviation Alternative Fuels Initiative (CAAFI®)—CAAFI® was jointly founded in 2006 by FAA and aviation industry associations A4A, AIA, and ACI-NA to explore the potential of AJFs. It has become a key forum for exchange, coordination, and cooperation among a cross section of participants from airlines, manufacturers, airports, fuel producers, Federal agencies, and international players and has been crucial in accelerating fuel approvals, critical research, and commercial agreements between fuel producers and airline fuel users.

---

[19] The Defense Production Act of 1950, As Amended, 50 U.S.C. App. § 2061 et seq. (2009), http://www.acq.osd.mil/mibp/dpac/final__defense_production_act_091030.pdf.

Appendix 3.
Federal Alternative Jet Fuel (AJF) Research and Development (R&D) Goals and Objectives

The following research and development (R&D) goals and objectives are intended to guide Federal R&D efforts in addressing key scientific and technical challenges. These R&D goals and objectives are projected to be completed over near-term (<5 years), mid-term (5–10 years), and far-term (>10 years) time horizons. Agencies that have an R&D mission that aligns with a particular goal or objective are indicated. In most cases, the goals and objectives are shared by multiple agencies, which reflects the reality that agencies have distinct expertise and capability and play complementary roles that contribute to overall AJF R&D enterprise. R&D Coordination takes place via bodies indicated in Appendix 2, including the Biomass R&D Board, CAAFI®, Farm to Fly 2.0, and so forth.

R&D Goals and Objectives for Feedstock Development, Production, and Logistics

Feedstock Development, Production, and Logistics Goals	Objectives		
	Near-Term (<5 years)	Mid-Term (5–10 years)	Far Term (>10 years)
Feedstock Development **(FG1) Increase crop yields (tons/acre), water and nutrient use efficiency, as well as pest and disease resistance, and improve feedstock conversion characteristics (DOC, DOE, NSF, USDA)**	(FG1NO1) Benchmark readiness of existing regional feedstock (Feedstock Readiness Level (FSRL)) and identify new and diverse feedstocks with regard to potential delivery at requisite quantity, quality, and cost for extant or emerging conversion platforms (DOE, USDA) (FG1NO2) Identify public and private sources for regional feedstock candidates and catalog characteristics to understand where research gaps exist (DOE, NSF, USDA) (FG1NO3) Set up feedstock improvement programs/ partnerships to facilitate accelerated crop improvement (DOE, USDA) (FG1NO4) Leverage existing feedstock improvement programs and genetic/ genomic information (DOE, USDA) (FG1NO5) Evaluate/ characterize agriculture/ forest residuals (DOE, USDA) (FG1NO6) Develop risk management tools such as crop insurance to promote dedicated bioenergy crop production (USDA)	(FG1MO1) Identify new traits/genes of interest and develop genomic and marker-aided selection capacities (DOC, DOE, NSF, USDA) (FG1MO2) Scale-up next-generation improved feedstock for deployment and large-scale site evaluation studies (USDA) (FG1MO3) Develop research/science-based process to move transgenic crops through the regulation process and implement the process (DOE, USDA) (FG1MO4) Evaluate and improve regionally and seasonally diverse third-generation crops (DOE, USDA) (FG1MO5) Develop efficient non-traditional crops, such as algae, microbial resources, or synthetic biology systems, that can use non-potable water or waste carbon monoxide/ carbon dioxide (CO/CO_2), and that can process/treat various industrial and municipal wastes as energy sources (DOE, NSF, USDA)	(FG1LO1) Scale-up improved next-generation feedstock for deployment and large-scale site evaluation studies (USDA) (FG1LO2) Scale-up transgenic crops that have been cleared for environmental release for deployment and large-scale site evaluation studies (DOE, USDA) (FG1LO3) Evaluate and improve regionally and seasonally diverse fourth-generation crops (DOE, USDA)

Feedstock Development, Production, and Logistics Goals	Objectives		
	Near-Term (<5 years)	Mid-Term (5–10 years)	Far Term (>10 years)
Feedstock Production **(FG2) Develop sustainable feedstock production systems that require minimal inputs, have a high tolerance for environmental stress, and minimize the risk of adverse environmental impacts (e.g., invasiveness, erosion) (DOE, USDA)**	(FG2NO1) Identify/evaluate/ leverage past/current regional R&D to understand gaps/needs (DOE, USDA) (FG2NO2) Leverage/build regional R&D partnerships including local communities, universities, industry, government, and non-governmental organizations (NGOs) (DOE, USDA) (FG2NO3) Set up/leverage sub-commercial-scale test sites to perform R&D with existing genotypes/residuals (DOE, USDA) (FG2NO4) Train regional extension/tech transfer specialists who will interface with producers (farmers/ forest landowners), processors, and communities (USDA) (FG2NO5) Identify regional workforce requirements across supply chains (USDA)	(FG2MO1) Evaluate regional feedstock production R&D to determine best practices, best regional feedstock/production systems/conversion platform pairings (DOE, USDA) (FG2MO2) Evaluate cropping/management system economic, environmental, and social sustainability criteria and indicators (DOE, USDA) (FG2MO3) Train additional regional extension/ technology transfer specialists who will interface with producers (farmers and landowners), processors, and communities (USDA) (FG2MO4) Evaluate/modify workforce development needs and training (USDA) (FG2MO5) Evaluate policy and economic incentives to promote sustainable, efficient production of dedicated bioenergy feedstock (DOE, USDA)	(FG2LO1) Evaluate regional feedstock production R&D to recommend long-term best practices, best regional feedstock/production systems/conversion platform pairings (DOE, USDA) (FG2LO2) Evaluate cropping/ management system economic, environmental, and social sustainability criteria and indicators for long-term analysis (DOE, USDA) (FG2LO3) Train additional regional extension/tech transfer specialists who will interface with producers (farmers/landowners), processors, and communities over the long term (USDA)

Feedstock Development, Production, and Logistics Goals	Objectives		
	Near-Term (<5 years)	Mid-Term (5–10 years)	Far Term (>10 years)
Feedstock Logistics **(FG3) Improve harvesting, collection, storage, densification, pretreatment, and transportation of physical biomass to the conversion facility (DOE, NSF, USDA)**	(FG3NO1) Identify/evaluate/leverage past/current regional R&D to understand gaps/needs (DOE, USDA) (FG3NO2) Study the impact of changes in regional logistic scenarios on the quality of existing feedstock/residue and bio-based product/ co-product options (DOE, NSF, USDA) (FG3NO3) Identify targets for reducing costs, minimizing feedstock losses, and enhancing quality at every step of the regional logistic chain (DOE, USDA) (FG3NO4) Establish/leverage feedback loops from regional AJF conversion platforms (DOE, USDA)	(FG3MO1) Optimize regional logistic chains with extant/emerging feedstocks/residues and conversion technologies (DOE, USDA) (FG3MO2) Understand feedstock quality impacts from logistic chain on bio-based product/co-product options (USDA, DOE)	(FG3LO1) Review and optimize, as needed, regional logistic chains with extant/emerging crops/residues and conversion technologies (DOE, USDA) (FG3LO2) Optimize feedstock quality impacts from logistic chain on bio-based product/ co-product options (DOE, USDA)
(FSG4) Improve collection, storage, densification, pretreatment, and transportation of municipal solid waste to the conversion facility (DOE)	(FG4NO1) Identify/evaluate/ leverage past/current regional R&D to understand gaps/needs (DOE) (FG4NO2) Study the impact of changes in regional logistic scenarios on the quality of existing MSW feedstock (DOE) (FG4NO3) Identify targets for reducing costs, minimizing feedstock losses, and enhancing quality at every step of the regional logistic chain (DOE)	(FG4MO1) Optimize regional logistic chains for MSW with emerging/extant conversion technologies (DOE) (FG4MO2) Understand feedstock quality impacts from logistic chain on co-product options (DOE)	(FG4LO1) Review and optimize, as needed, regional logistic chains for MSW with extant/ emerging conversion technologies (DOE) (FG4LO2) Optimize feedstock quality impacts from logistic chain on co-product options (DOE)

R&D Goals and Objectives for Fuel Conversion and Scale-Up

R&D Goals	Objectives		
	Near-Term (<5 years)	Mid-Term (5–10 years)	Far Term (>10 years)
(S1) Enable discovery, development, enhancement, and scale-up of conversion processes with improved yield, efficiency, and energy requirements that lead to cost-competitive AJF (DOD, DOE, NSF, USDA)	(S1NO1) Improve conversion yields for promising processes by increasing the lifetime of catalysts, removing oxygen efficiently, and producing fuel precursors that can be easily converted to jet fuel (DOE, NSF) (S1NO2) Develop new and effective pre-treatment technologies that improve the ability of biomass to be converted to AJF, including co-processing in existing petro-based refineries (DOE, NSF) (S1NO3) Develop and demonstrate integrated fuel conversion facilities at commercial scale that produce jet fuel as a component of their product slates (DOD, DOE, USDA)	(S1MO1) Develop new approaches that permit reduced numbers of operational steps (e.g., consolidated bio-processing) (DOE, NSF)	(S1LO1) Research and enable economically viable and sustainable processes that can use a variety of feedstocks to produce jet fuel and can be operated at a pilot/demonstration scale (DOE)
(S2) Develop conversion technologies that can produce jet fuel from multiple feedstocks in a distributed manner (DOD, DOE, NSF, USDA)	(S2NO1) Determine the feasibility of co-feeding biomass, waste, and other feedstocks for conversion process at demo and pilot scales (DOE)	(S2MO1) Enable the use of stranded resources (natural gas, biomass, and other feedstocks currently being flared or wasted) for the production of jet fuel (DOE)	(S2LO1) Develop and demonstrate technologies based on biological, chemical, or catalytic principles that operate efficiently at distributed scale (DOD, DOE, NSF, USDA)

R&D Goals and Objectives for Fuel Testing and Evaluation

R&D Goals	Objectives		
	Near-Term (<5 years)	Mid-Term (5–10 years)	Far Term (>10 years)
(Q1) Facilitate civil and military approval of additional AJF pathways by enabling efficient evaluation for performance and safety through advancement of certification and qualification processes and collection and analysis of data (DOD, DOT, NASA)	(Q1NO1) Support capability to perform all testing required by ASTM D4054[20] and military specifications, including fuel property testing, component/rig testing, and aircraft engine testing to complete evaluation for viable alternative jet fuel pathways (DOD, DOT) (Q1NO2) Establish a coordinated process to track and monitor progress of ASTM and military AJF task forces, conduct data review and testing activities, and establish schedules and prioritize projects leveraging tools such as the Fuel Readiness Level (FRL) (DOT) (Q1NO3) Explore novel approaches for approval of alternative jet fuels with test and evaluation requirements commensurate with the blend percentages (DOD, DOT) (Q1NO4) Characterize the conventional U.S. jet fuel supply to better understand the jet fuel property baseline and variations to enable statistical assessment of alternative jet fuel appropriateness and benefits (DOT) (Q1NO5) Advance fuel composition and combustion performance modeling, experimentation, and analysis (DOD, DOT, NASA) (Q1NO6) Develop improved	(Q1MO1) Integrate fuel composition and combustion modeling and tests into the alternative jet fuel approval process to support broad-based approvals including neat and blended fuels (DOD, DOT, NASA)	(Q1LO1) Develop high-speed, efficient, affordable test methods that can be integrated into the alternative fuel production and distribution infrastructure to allow real-time compositional verification of alternative jet fuel properties to reduce or eliminate the need for process-specific qualification (DOD, DOT, NASA)

---

[20] ASTM International, *Standard Practice for Qualification and Approval of New Aviation Turbine Fuels and Fuel Additives*, ASTM D4054-14 (West Conshohocken, PA: ASTM International, 2014).

R&D Goals	Objectives		
	Near-Term **(<5 years)**	**Mid-Term** **(5–10 years)**	**Far Term** **(>10 years)**
	test methods for certification and qualification that allow for more rapid, efficient and less costly fuel evaluation to support approvals (DOD, DOT, NASA)		
(Q2) Improve scientific understanding of how AJF composition impacts gas turbine combustion emissions and operability (DOC, DOD, DOT, NASA)	(Q2NO1) Establish national databases of alternative jet fuel engine operability testing and combustion emissions data and analyses (DOC, DOT) (Q2NO2) Support AJF turbine engine combustion emissions measurements and analysis (DOD, DOT, NASA) (Q2NO3) Examine dependence of variations in jet fuel composition on magnitude and types of combustion emissions (DOD, DOT, NASA)	(Q2MO1) Expand AJF turbine engine combustion emissions measurements and analysis to include new AJF candidates (DOD, DOT, NASA) (Q2MO2) Expand national database of combustion emissions measurements and analyses to include data for new AJF candidates (DOC, DOD, DOT, NASA) (Q2MO3) Develop correlation and parametric relationship between jet fuels combustion and combustion emissions (DOC, DOD, DOT, NASA)	(Q2LO1) Develop modeling capability to predict jet engine combustion emissions as a function of fuel composition (DOD, DOT, NASA) (Q2LO2) Employ fuel composition characteristics to improve combustor operability and enable combustor designs that can accommodate a broader range of fuel properties (DOD, NASA)

R&D Goals and Objectives for Integrated Challenges

R&D Goals	Objectives		
	Near-Term (<5 years)	**Mid-Term (5–10 years)**	**Far Term (>10 years)**
(C1) Advance understanding of and improve environmental sustainability of AJF production and use (DOC, DOT, EPA, USDA)	(C1NO1) Advance the scientific understanding of environmental impacts of AJF production and use on all relevant scales, including those related to life-cycle emissions that impact climate change and environment (DOC, DOT, EPA, USDA) (C1NO2) Improve capabilities to assess natural resource requirements for AJF production and use on regional and national scales (DOC, DOT, USDA) (C1NO3) Compile, assess, and disseminate definitions, protocols, data, and tools in support of environmental sustainability analysis (DOC, DOT, EPA, NSF)	(C1MO1) Contribute to internationally recognized approaches to AJF environmental sustainability assessment through quantitative sustainability analysis (DOC, DOT, EPA, USDA) (C1MO2) Develop best practices for collecting and analyzing AJF life-cycle inventory data aligned with each stage of fuel and feedstock readiness (DOE, DOT, EPA)	(C1LO1) Develop improved tools and approaches for environmental sustainability assessment applicable across all stages of the supply chain and generalizable to all AJF pathways (DOD, DOT, EPA, USDA)
(C2) Develop and validate a comprehensive systems model to support viable AJF deployment (DOD, DOE, DOT, EPA, USDA)	(C2NO1) Identify and quantify the elements and interactions among elements that are required to develop a systems model that can be used to create regional and national scenarios for AJF deployment that reflect criteria for environmental, economic and social sustainability (DOT, EPA, USDA) (C2NO2) Advance techno-economic and regional development path analyses of AJF (DOE, DOT, EPA) (C2NO3) Evaluate impacts of AJF production on social sustainability (DOE, DOT, USDA)	(C2MO1) Develop analytical capabilities for components of the systems model (DOC, DOE, DOT, USDA) (C2MO2) Develop an integrated framework that combines economic, environmental, and social models to assess AJF production and use within and across regions (DOC, DOE, DOT, USDA) (C2MO3) Create and validate a systems model for regional AJF production supporting plug-and-play engineering decision-making (DOD, DOE, NSF)	(C2LO1) Implement a comprehensive systems model to examine scenarios and identify potentials for improved sustainability of AJF production and use enabling plug-and-play engineering algorithms (DOE, DOT, EPA, USDA)

R&D Goals	Objectives		
	Near-Term (<5 years)	Mid-Term (5–10 years)	Far Term (>10 years)
(C3) Promote communication as well as scientific and technical R&D best practices for the national enterprise (DOC, DOE, DOT, USDA)	(C3NO1) Facilitate the dissemination of scientific and technical information via improved access to outcomes of federally funded AJF R&D activities (DOC, DOE, DOT) (C3NO2) Identify new areas for scientific research that cuts across individual components of the development path (DOE, DOT, NSF)	(C3MO1) Implement meta-analysis to evaluate the state of knowledge and identify research challenges for the future (DOE, DOT, USDA)	

References

Agricultural Act of 2014, H.R. 2642, 113th Cong. (2014), https://www.gpo.gov/fdsys/pkg/BILLS-113hr2642enr/pdf/BILLS-113hr2642enr.pdf.

ASTM International, Standard Practice for Qualification and Approval of New Aviation Turbine Fuels and Fuel Additives, ASTM D4054-14 (West Conshohocken, PA: ASTM International, 2014).

Biomass Research and Development Board (BR&DS), Federal Activities Report on the Bioeconomy, briefing (Washington, DC: Department of Energy, Bioenergy Technologies Office, March 8, 2016), http://biomassboard.gov/pdfs/tac_2016_q1_baumes.pdf.

Biomass Research and Development, "Board," last updated 03/16/2015, http://www.biomassboard.gov/board/board.html.

Biomass Research and Development, "Initiative," last updated 12/09/2010, http://www.biomassboard.gov/initiative/initiative.html.

Biomass Research and Development, "Advisory Committee," last updated 03/16/2015, http://www.biomassboard.gov/committee/committee.html.

Defense Production Act of 1950, As Amended, 50 U.S.C. App. § 2061 et seq. (2009), http://www.acq.osd.mil/mibp/dpac/final__defense_production_act_091030.pdf.

Department of Defense, "DoD Energy Policy," Department of Defense Directive (DODD) 4180.01 (Washington, DC: USD (AT&L), April 16, 2014), 1, http://www.dtic.mil/whs/directives/corres/pdf/418001_2014.pdf.

Food Conservation and Energy Act of 2008, Pub. L. 110-246, 121 Stat. 2064 (2008), https://www.gpo.gov/fdsys/pkg/PLAW-110publ246/pdf/PLAW-110publ246.pdf.

National Science and Technology Council, National Aeronautics Research and Development Plan (Washington, DC: Executive Office of the President, February 2010), https://www.whitehouse.gov/sites/default/files/microsites/ostp/aero-rdplan-2010.pdf.

Office of Energy Efficiency & Renewable Energy, "Biomass Research and Development Act of 2000," http://energy.gov/eere/bioenergy/downloads/biomass-research-and-development-act-2000.

Regulation of Fuels and Fuel Additives: Changes to Renewable Fuel Standard Program; Final Rule, 75 Fed. Reg. 14670 (March 26, 2010), https://www.gpo.gov/fdsys/pkg/FR-2010-03-26/pdf/2010-3851.pdf.

Regulation of Fuels and Fuel Additives: Identification of Additional Qualifying Renewable Fuel Pathways under the Renewable Fuel Standard Program, 78 Fed. Reg. 14190 (March 5, 2013), https://www.gpo.gov/fdsys/pkg/FR-2013-03-05/pdf/2013-04929.pdf.

White House, *Blueprint for a Secure Energy Future* (Washington, DC: Executive Office of the President, March 2011), http://www.whitehouse.gov/sites/default/files/blueprint_secure_energy_future.pdf.

White House, *National Bioeconomy Blueprint* (Washington, DC: Executive Office of the President, April 2012), http://www.whitehouse.gov/sites/default/files/microsites/ostp/national_bioeconomy_blueprint_april_2012.pdf.

White House, *The President's Climate Action Plan* (Washington, DC: Executive Office of the President, June 2013),
https://www.whitehouse.gov/sites/default/files/image/president27sclimateactionplan.pdf.

Abbreviations

A4A	Airlines for America
ACI-NA	Airports Council International-North America
AFRI	Agriculture and Food Research Initiative
AIA	Aerospace Industries Association
Aireg	Aviation Initiative for Renewable Energy in Germany
AISAF	Australian Initiative for Sustainable Aviation Fuels
AJF	alternative jet fuel
AJF-IWG	Alternative Jet Fuel Interagency Working Group
ARPA-E	Advanced Research Projects Agency-Energy
ARS	Agricultural Research Service
ASCENT	Aviation Sustainability Center
ATJ	alcohol-to-jet
BBP	Brazilian Biofuels Platform
CAAFI®	Commercial Aviation Alternative Fuels Initiative
CAEP	Committee on Aviation Environmental Protection
CCC	Commodity Credit Corporation
CLEEN	Continuous Lower Energy, Emissions, and Noise
CO	carbon monoxide
CO2	carbon dioxide
DLA	Defense Logistics Agency
DNA	deoxyribonucleic acid
DOC	Department of Commerce
DOD	Department of Defense
DODD	Department of Defense Directive
DOE	Department of Energy
DOI	Department of Interior
DOT	Department of Transportation
DPA	Defense Production Act
EPA	Environmental Protection Agency
FAA	Federal Aviation Administration
FRL	Fuel Readiness Level

FSRL	Feedstock Readiness Level
GAMA	General Aviation Manufacturers Association
GHG	greenhouse gas
ICAO	International Civil Aviation Organization
ITA	International Trade Administration
MOU	Memorandum of Understanding
MSW	municipal solid waste
NASA	National Aeronautics and Space Administration
NBAA	National Business Aviation Association
NGO	non-governmental organization
NIFA	National Institute of Food and Agriculture
NIST	National Institute of Standards and Technology
NOAA	National Oceanic and Atmospheric Administration
NOX	nitrogen oxides
NSF	National Science Foundation
OEPNU	Office of Energy Policy and New Uses
OSTP	Office of Science and Technology Policy
PGRP	Plant Genome Research Program
R&D	research and development
RFS	Renewable Fuel Standard
RIN	Renewable Identification Number
SBIR	Small Business Innovation Research
SOX	sulfur oxides
TAC	Technical Advisory Committee
U.S.C.	United States Code
USDA	U.S. Department of Agriculture